BATTERY

For Energy Storage
Essential Research

FOUZIA BEGUM

ISBN: Softcover   978-1-5434-8039-9
         EBook   978-1-5434-8038-2

Print information available on the last page

Rev. date: 02/12/2018

To order additional copies of this book, contact:
Xlibris
1-888-795-4274
www.Xlibris.com
Orders@Xlibris.com

CONTENTS

Dedicated to all Efficient energy

Storage device known as Smart Battery

FOREWORD

Tech Insider, Drake Bear recently interviewed Bill Gates where the Microsoft giant discusses the task to bring electricity to the billions of energy deprived people while finding an eco-friendly power source that don't kill the earth. Bill Gates acknowledged that storing energy is a hard task because no miraculous storage solution is available on hand yet or in the market. Assessing the market quiet intently he provided answers to question spanning a bunch of energy issues. For example, for improvement question he says: "In fact, batteries haven't improved over the last 100 years as much as they would need to in order to make that happen". This suggests a stagnant situation crumbles in battery technology that is hindering progress. Energy innovation, efficient electronics use and conservation of energy topics are three interrelated branches of one primitive issue named "energy in life." He continued; "to bring a change or progress here level of all three braches need to be same. Lack of progress in any branch bring a status quo situation. That's what happening in third world countries now. People do not feel any change there because the mismatched levels branch'. Bill Gate has invested now a lot of money in battery companies — and he thinks that there's a lot of battery companies exists here and there about which he is not aware. But his assumption is that all companies are having a tough time achieving a quick solution for storing energy. According to Bill Gate, "All paths needs to revise to find a better solution, things like the wind in the jet stream, which is very high up. The material science of what type of kite string you would need to connect up to that. That's still at the basic research level." Read: Full interview link is here.

PREFACE

Katherine Hamilton, policy director of Energy Storage Association, the industry's leading trade group in USA evaluated recently Tech development using a different prism. In contrast to Bill Gate, Katherine has seen huge progress there which is why she commented as saying, "Energy storage is kind of like pizza from the freezer. It's there when you want it, and you don't have to wait for it." Her valuation of energy storage with Pizza for 'to Bake and Serve' – anywhere, anytime though a little exaggerated, yet it represent a very technically advanced society. If stored energy chunks be available in the market, be ready to use with a mouse click – transform rapidly to a mode to run electric vehicles or hand held gadgets then certainly this is a technical jump. And, Katherine Hamilton is not the only person there to boast energy chunks' influence on civilized life, rather many people out there in many places consider in line with her. So, the real question now is: why is energy solvency needed anyway? The most persuasive answer: it is needed, because it allows people to budget their energy consumption.

Consumers know batteries need to be replaced at exact time for devices that thrives on battery power. A rechargeable, long lasting and a cheaper product's name remains on demand when replacement for battery is needed. Yet, customers do get confuse because so many 'Brands' available in market now! Which one is better?

Technology has evolved rapidly in all branches these days – opportunities to upgrade a storage device to introduce a better product has been accomplished partially. Today, lithium-ion battery, (LIB) has been introduced everywhere from smart phone to hybrid Electric Vehicles' (EVs). LIB's 'electron density – found to be high and it charges faster. So, LIB is providing power efficiently satisfying customers' demands. By manipulating LIB's lightweight property, electric vehicle manufacturers are expecting to deliver important breakthroughs in electric transportation now.

Core purpose of composing this book is to provide appropriate energy storage tips, provide answers for unattended questions, provide better energy storage solution to make public's life easier. I encourage customers from all walks of life to buy this book and get to watch "online" videos which are helpful for life and energy. I discussed methods in this book for public to be able to get aware about energy storage issues. To run households and residential solar power devices using a powerful battery with steady supply of electricity is still perplexing. In presenting this book we simplified all complex terms, rules or terms related with battery, so that common curious people feel at ease and switch 'brand to brand.' I advocate for "Free Solar Energy" —Learn here now how to convert and store Free Solar Energy. Read this book at- http://www.fbsolarllc.com.

And subscribe.

CHAPTER-1

INTRODUCTION

Till date battery researchers characterized three essential factors to upgrade: i) to reduce charging time, ii) to reduce the frequency of charging, iii) and the overall cost of a new storage device. Cutting down charging time and cutting down frequency of charge will transform the energy cell to an efficient one. We do understand the 'cost term. With these change—a "smart battery" develops. For every cell; Rule of thumb is: to be commercially viable all energy industry batteries need to be 99.9% efficient.

Our sweet homes require electrical appliances all the time—phones, laptops, plug-in vehicles, electric bikes, kitchen appliances, portable computers, clocks or watches including children's toys. Driving force for building battery systems is demand for charges -- when total electricity consumption hits or exceeds there. Often, we install hydropower, geothermal, solar or, wind power plants at our backyard with variable renewable sources of power to produce electricity at certain times of day. By installing a powerful energy storage system near the plant, literally we can store all the renewable power (solar, in particular) into a battery that provides power even at night. Some battery provides backup power. The net point is, by installing efficient (smart) battery to store electric energy is one practical solution to face energy shortage situation.

Examples:

- TESLA's Powerwall battery system draws power either from solar panels or from the electric grid and simply store unused power as "frozen pizza." It is understood that when the grid supply disappears the battery takes over and discharges for ten, twelve, or for longer hours. Some batteries can charge 3 times faster than it takes to discharge.

- Power Grid is when it functions we can grab the available electrons as efficiently as we can and store it.

- Alternative to power grid is electrochemical solution, is the other best option. Here, we look for, Redox-Flow Vanadium storage system technology found to be handy- its huge cell size, easy to operate to power a home like a grid-based energy storage system. Watch fun but useful <u>videos from the link</u> here:

Other than this, regular everyday batteries that are available in the markets today as **primary or secondary** cells, Lead-acid battery, Daniel cell, redox-flow battery, Nickel-Hydride (NiMH) etc., each type has unique voltage variation from <1.0 V to – 10 .0V> and with specific power densities (PD). Most recent addition in the market-shelves are lithium-ion batteries that rely on a chemical reaction of Lithium metal releasing electrons which is then extracted into an electric circuit.

To design Electric Vehicle (EV) for transportation, technologies modifies battery design for storage all the time. EV requires a high **power density (PD)** battery.

CHAPTER-2

NOTES ON GENERAL BATTERY PARTS

<u>Battery Cell:</u> A battery is an electrochemical cell. The core unit of battery that appears in all cells sketched here [Sketch 1.1.,] followed by various cell definitions. Any new battery construction means redesigning the core unit or, by bringing modifications on the electrodes, or, in electrolyte used. A change in any part of the core unit will generate a new cell with new voltmeter reading (V). Voltage difference between two electrodes is the driving force of a battery.

Core Battery Unit

A battery is an **electrochemical cell**. All batteries have three basic components: an electrolyte to provide electrons, an anode to discharge those electrons, and a cathode to receive the electrons.

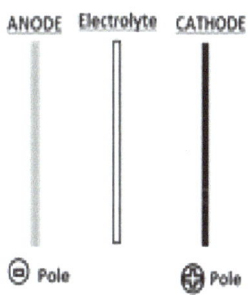

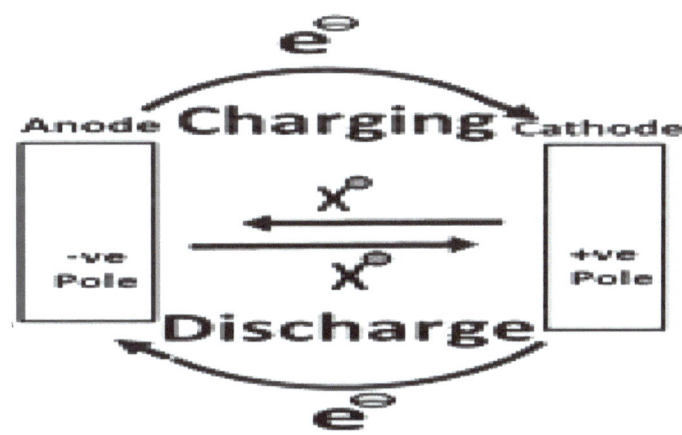

1.1.1 SKETCH: CORE BATTERY UNIT (TOP), CHARGEDISCHARGE DIRECTIONS (DOWN)

Galvanic cells or, Voltaic cells or, Electrochemical cells are driven by a spontaneous chemical reaction (see ECE link) that produces an electric current using outside circuit. These cells are the basis for the batteries that fuel modern society and are therefore important.

Electrochemical cell is a device capable of either generating electrical energy from chemical reactions or facilitating chemical reactions through the introduction of electrical energy. An example of an electrochemical cell is a standard **1.5-volt battery**, which is a **primary cell** with chemical composition of carbon, zinc and Ammonium chloride.

Rechargeable batteries are **secondary cells**, consist of reversible cell reactions that allow them to recharge, or regain their cell potential, through the work done by passing currents of electricity. A rechargeable battery can charge and discharge numerous times. For example; lithium-ion battery known as LIB.

Daniel cell is an electrochemical cell (named after John Frederic Daniell, a British chemist) consisted of a copper pot filled with copper sulfate solution dipped inside of which was a porous ceramic pot filled with sulfuric acid and a zinc electrode.

Daniel Cell (E^0 = 1.10V, #of electrons in play = 2)

Spontaneous electrochemical reactions involve transfer of electrons.

One species loses electron, $X^0 \xrightarrow{-e^-} X^+$ Oxidation

One species gain electron, $Y^+ + e^- \longrightarrow Y^0$ Reduction

To get current, the two halves (Oxidation II Reduction) of the chemical reactions must be separated. The halve cells are then connected through appropriate wiring like a galvanic cell. The cell to the right is a Standard Daniel cell. Cell notation:

$$(Cu|Cu(SO_4)_2||ZnSO_4|Zn)$$

Overall reaction: $Cu^{2+}_{(aq)} + Zn^0_{(s)} \longrightarrow Zn^{2+}_{(aq)} + Cu^0_{(s)}$

$$E^0_{cell} = E^0_{red} - E^0_{ox} = \text{Voltage in open circuit} = 1.10V$$

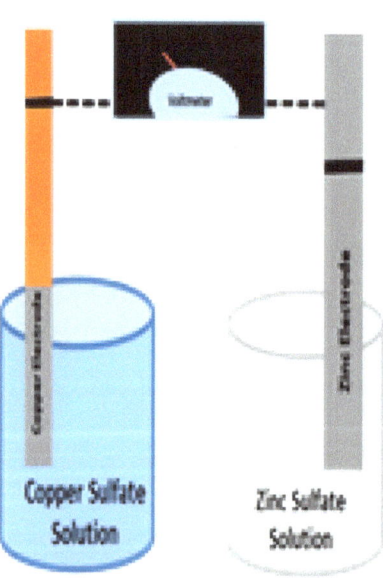

Copper Sulfate Solution   Zinc Sulfate Solution

1.1.2 DANIELCELL WITH CORE BATTERY FEATURES

A brand new cell, varying price, power or property as an electron density etc. always demonstrates core battery features that remain identical and follow everywhere.

2.1 RESEARCH REPORTS (CELL)

1. Aquino Energy, is the provider of **Aqueous Hybrid Ion (AHI)** battery systems. Aqueous Hybrid Ion (AHI) chemistry consists of a saltwater electrolyte, manganese oxide cathode, carbon composite anode, and synthetic cotton separator. This battery uses noncorrosive reactions at anode and cathode to prevent deterioration of the materials. This water-based chemistry results in a nontoxic and noncombustible product that is in abundance and safe to handle and environmentally friendly. It is of low-cost and are designed for long-duration cycle applications. AHI's new energy storage technology is more viable, especially in remote locations. The batteries are now optimized for long charge and discharge cycles for solar and other renewable energy storage applications.

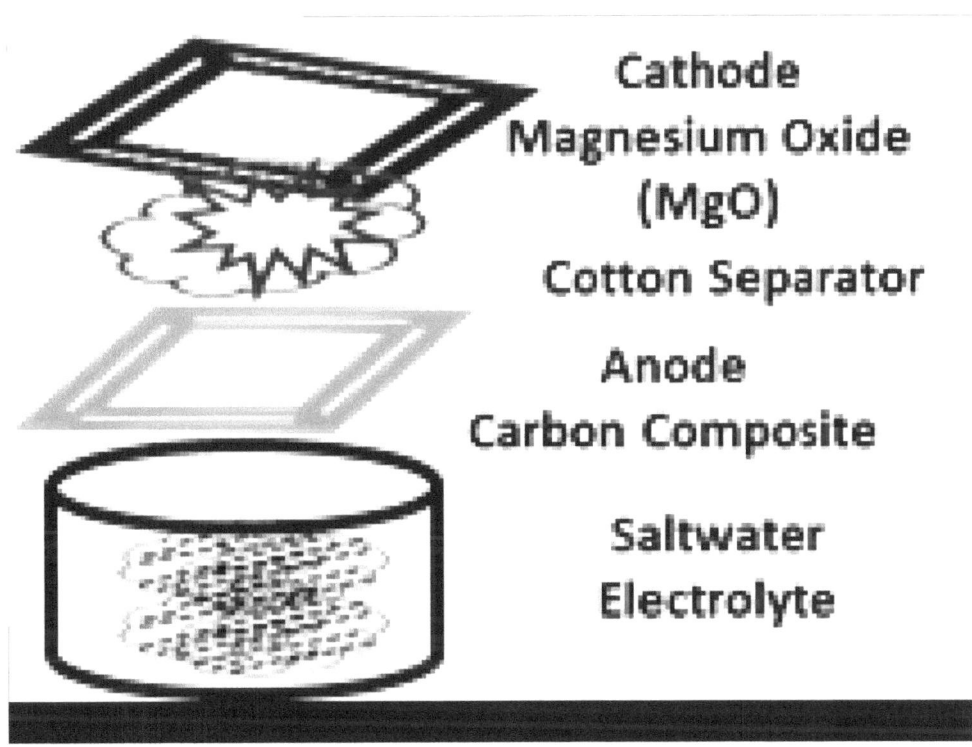

1.1.3 SKETCH AQUION (AHI) BATTERY

For—storing and transporting renewable power is cumbersome. Greensmith CEO John Jung singled out four points: batteries from Aquino uses aqueous sodium-ion chemistry, provides energy efficiency **85 %**; a multi-hour energy storage capacity and a price point of $250 **per kilowatt-hour (KWh)**.

2. British company [1(video)] **Isentropic** has developed a system that could store energy in cells that are filled with gravel. Using Argon gas and a heat-exchange pump, Electrical energy from solar panels or wind turbines is used to warm the gravel. When the energy is required, the process is reversed, and the heat is used to drive an engine that creates power ready for the grid. This system considered inefficient because it returns only 75 percent of energy

that we put in. The other disadvantage is the need of a tank full of Argon for the gravel to dip in. But, low cost and ease of installation is the core attraction. Energy for 8hrs only.

3. Liquid Metal Battery (LMB)

Donald Sadoway is the inventor of [2(video)] Liquid Metal Batteries (LMB). LMB brought changes slowly that helps lowering risk in the energy industry.

LMB project developed a low cost but long lifespan battery with grid-scale stationary energy storage. The battery is built upon three liquid layers as electroactive components; these consists of a liquid metal positive electrode, a fused salt electrolyte, and a liquid metal negative electrode. The three liquid layers float on top of one another based to their density differences and immiscibility. This assures low assembling cost and the use of inexpensive materials. Additionally, use of liquid electrodes to run the battery reaction mechanism smooth and long lasting. Sadoway's liquid-metal batteries provides affordable and efficient power storage. Laptops, cellphones etc. use identical grid-scale battery technology. Cambridge Company now known as Ambri Inc, is first to be built large scale LMB in New Mexico in USA, and in China and South Africa.

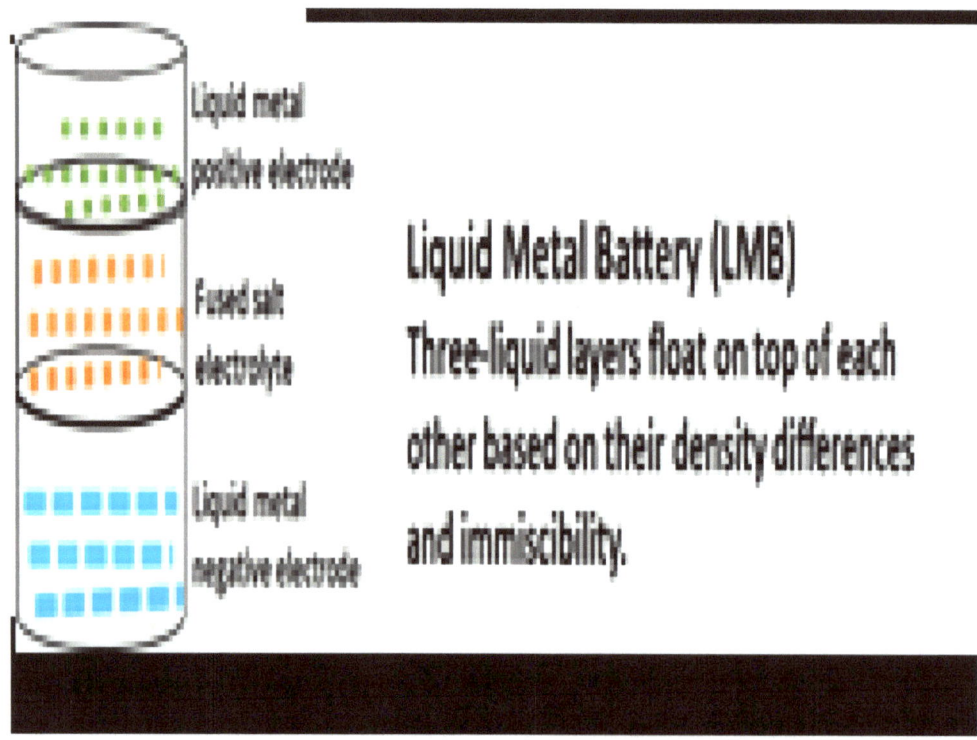

1.1.4 SKETCH LIQUID METAL BATTERY (LMB)

4. DSSC

Researchers at Ohio State University (OSU) have invented a solar battery – where they integrated two electric functions into one device (a solar cell+ a battery). The design is for to reduce the cost but for renewable energy propagation objective. The solar battery part consists of three electrodes: a Li-metal anode, an oxygen electrode made on carbon paper and a photoelectrode.

The discharge process remains identical as that of a conventional Li–O$_2$ (lithium superoxide) battery with the formation of Li$_2$O$_2$ (lithium peroxide) at the oxygen electrode. The charging process is little different.

Cell construction: Anode is the lithium plate (bottom), top of which lies a set with the composition: electrolyte — thin porous carbon sheet — electrolyte, on top of this set lies a permeable titanium gauze mesh containing a dye-sensitive TiO$_2$ photoelectrode. Photoelectrode generates triiodide ions (I$_3^-$) under illumination that spread on to the oxygen electrode surface via an iodide "shuttle." Here, they then oxidize into lithium peroxide.

Electrons within the constructed cell chemically decompose lithium peroxide into lithium ions and oxygen, with the oxygen released into the air and lithium ions stored as lithium metal. When the battery discharges, it pulls oxygen from the surrounding environment and consumes it to re-form lithium peroxide – after which the cycle can repeat again.

Permeable titanium gauze mesh contains dye-sensitive TiO$_2$ photoelectrode

Layer of Electrolyte (O$_2$)
Thin sheet of porous Carbon $\mathcal{C}$
Layer of electrolyte(O$_2$)

Anode -Lithium plate

$$2Li^+ + O_2 \underset{\text{Charging +2e}}{\overset{\text{+ 2e- Discharge}}{\rightleftharpoons}} Li\text{-}O\text{-}O\text{-}Li \quad \text{Lithium peroxide}$$

$$4Li^0 \underset{\text{reduction +4e}}{\overset{\text{- 4e oxidation}}{\rightleftharpoons}} 4Li^+$$

$$4Li^+ + 2O^{2-} \longrightarrow 2Li_2O \quad \text{Lithium oxide}$$

$$O_{2(gas)} + 4e^- \underset{\text{reduction}}{\longrightarrow} 2O^{2-}$$

1.1.5: CONSTRUCTION OF DSSC

When plugged, electrons in the battery chemically decompose lithium peroxide into lithium ions (LI$^+$) and oxygen O$_2$), with the oxygen released into the air and lithium ions stored as lithium metal. When the battery discharges, it pulls oxygen from the surrounding environment and consumes it to lithium peroxide – after which the cycle can repeat again.

5. 12V Car Battery: Lead Acid

Lead acid battery of which heavy lead plates and acid electrolyte (0.3M HeSO$_4$) are in use, is being replaced currently with lithium ion battery as in TESLA's model S.

AGM technology became popular in the early 1980s developed a sealed lead acid battery to reduce weight and improve reliability. The acid is absorbed by a very fine fiberglass mat, making the battery spill-proof also known as enhanced lead-acid battery. This enables shipment without hazardous material restrictions. AGM has very low internal resistance, is capable to deliver high currents on demand with a low self-discharge, offers a relatively long service life, and responds to low temperatures. Heavy metal lead is toxic causing the batteries' to decline.

[Chem. Rxn: Pb$_{(metal)}$ + H$_2$SO$_{4(acid)}$ → Pb(SO$_4$) + H$_{2(gas)}$]

6. Johnson Controls (JCI)

Is the world's largest maker of lead-acid 12-Volt starter batteries. Toshiba developed **Johnson Control Inc. (JCI)** battery center offering chemistry that recharges quickly, and works well in a wide range of temperatures. JC introduced new 12-V Lithium Titanate 2013 Ram 1500 battery-is an engine start-stop system (has not been commercialized yet.)

The company has developed 48V compact battery pack to be fitted in a smaller shoebox size and redesigned to integrate into a car's 12-Volt electrical system as part of a new start-stop setup. The system consists of a 48-volt lithium-ion battery pack and a low-voltage lead-acid battery programed to work with regenerative braking and to support higher power loads. The lead battery is from AGM technology serve as power accessories. Johnson Controls believes that small batteries can have a big impact on fuel efficiency. Fact is: Smaller lithium-ion battery packs only have an expected life of four years whereas, full-size lithium-ion packs used in hybrids and plug-in cars shows 10 years of life. So, large lithium-ion packs for plug-in vehicles are better batteries that help make cars greener. Replacing a normal 12-volt starter battery, to the small lithium-ion packs or, to large lithium-ion pack all modification is doable but expensive effort is the opinion of JC.

In Southern India a similar 12V lead acid battery Industry with the brand name of **JC (Jaychandran)** operates to serve the energy demands for the entire country to and its neighbors.

7. Flow Batteries

Flow Batteries are low-cost, high performance energy storage. A flow battery is formed by two liquids with opposite charge (electrolytes) which turn chemical energy into electricity by exchanging ions through a membrane. The largest and most stable flow batteries are ones that can be charged thousands of times without suffering degradation or capacity loss: best possible with vanadium metal. Four types of Flow Battery ideas have been presented here; **BBC, ViZn, Gildemeister, and ZnPoly** all reported that vanadium can lose or gain electrons very easily and by exploiting this property, energy storage technologies advanced its objectives.

7.1 <u>Imergy Power Systems</u> is the 1st developer of vanadium flow batteries in the market. To date ESP 250 series is the largest line of flow battery from Imergy, fit to operate for 20 years or more (with no need to change or replacement of electrolytes), capable to deliver 250 kilowatts of electrical power for four or more hours. ESP 250 series is ideal for utilities, large commercial, industrial and government end-users, and renewable energy projects. The batteries are more cost-effective to meet peak electricity demand. Smaller-scale batteries are ESP5 and ESP30 vanadium-flow battery systems, an integrated, 'plug-and-play' platform.

<u>Chemistry of redox flow battery</u>: Vanadium chemistry of which oxidized vanadium in sulfuric acid— vanadium is stripped of its five outermost electrons--turns to a yellow solution. Through gradual change in color from green, blue and violet in presence of zinc, color change represents electrons being passed to the vanadium.

$$V^0 \xrightarrow[\text{Oxidation}]{H_2SO_4} V^{5+} + 5e^- \quad \Big| \quad V^{5+} + 5e^- \xrightarrow[\text{Reduction}]{Zn} V^0$$

Yellow solution   Green to Blue to Violet

Rxn:1-Vanadium (Oxidation-Reduction)

This process then applied to on much larger scale battery, and then it termed as Vanadium Redox Flow Battery; consists of two giant tanks of different solutions of Vanadium separated by a membrane. **KEY**: Redox-Fow-Battery:

- Current generates as fluids in tank is pumped on either side of the electrodes.

- Vanadium releases electrons in one tank turning solution color to yellow.

- Vanadium receives electrons in the other tank turning solution color from blue to green to violet.

- Electron movement generates current. Matched number of protons (H^+ ions) pass across the membrane between the solutions simulltaneously.

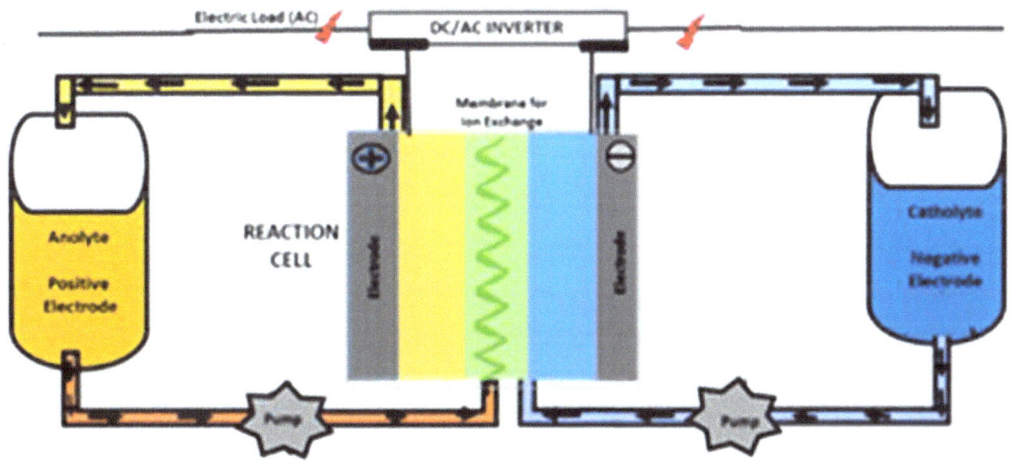

1.1.7 REDOX-FLOW BATTERY

Vanadium has long life and it is recyclable, can be charged and discharged 20,000 times without loss of performance--about ten times that of Li-ion batteries. The battery is nontoxic, safer and reliable:.

i) Cost- Refined Vanadium is too expensive.

ii) Huge size- only large scale projects fit in design. No reduction in size.

iii) Not suitable for Electric cars or garage, but effectively used by electrical utilities and other large-end users compatible with renewable, such as solar will be better.

iv) Energy Density of Redox-flow batteries is only a third of the energy density of a lithium battery (ED_{redox} = 0.33($ED_{Lithium}$).

7.2. **ViZn Energy Systems** is another energy storage facility that uses zinc redox flow battery technology. ViZn's system utilizes safe, nontoxic, nonexplosive zinc-iron electrolyte. The Company aims to produce 80-kilowatt-h and 160-kilowatthour system.

7.3 **Gildemeister Energy Solutions** introduced energy storage based on vanadium redox flow technology. Germany's Gildemeister sets up showroom like CellCube in USA offering 60-plus megawatts of storage facility in the market. System is independent of weather and temperature fluctuations.

7.4. **PNNL** (Pacific Northwest National Laboratory) researchers led by Wei Wang have developed a prototype of high-performance zinc-polyiodide flow battery with a high energy density of 167 Wh/l (watt-hours per liter) – and that remained open for optimizations. The high-energy density of aqueous zinc-polyiodide flow battery uses highly soluble iodide/triiodide redox couple, with a discharge energy density of 167 Wh/l using 5.0 M ZnI_2 electrolyte.

The finding is described in an open access of [5]Nature Communications paper.

8. Nickel-Metal-hydride (NiMH)

A nickel–metal hydride battery, abbreviated **NiMH** is a rechargeable battery. These batteries are used only in hybrid cars – at least those from Toyota. For use in plug-in cars of all sorts high energy density NIMH cells made it practical.BASF researchers boosted the energy density of NiMH cells recently and brought changes to the microstructure of nickel-metal electrodes, making it more durable for the same power production. NiMH cells current energy density is of 140 Wh/Kg. NiMH cells are safer than lithium ion batteries. Though, ED of NiMH is still down – considering two factors that LIB degrade over time at higher rates than do NiMH cells and the battery is considered to safer due to the lack of a liquid electrolyte causing any leak or spill related accidents.

Chemistry of NiMH: The chemical reaction with positive electrode which is from **nickel** oxyhydroxide (NiOOH), and negative electrodes uses hydrogen-absorbing alloy. NiMH battery's energy density is similar to that of a lithium-ion battery. Cost per kilowatt-hour now is > $150.

9. **Stanford's Aluminum Battery**

Stanford University researchers developed a new aluminum-ion battery that promises to deliver a number of improvements over the lithium-ion battery packs. This new Stanford's battery technology is less prone to explode, and it can also reach a full charge in just minute. The aluminum-ion battery model is extremely durable, and can endure a greater number of charge/recharge cycles (7,500 charging cycles without losing performance).

An excited but <u>unproven technology</u> that only exist now in research labs.

X-X

2.2 LITHIUM ION BATTERIES (ALL TYPES)

10. Lithium Battery: Lithium batteries have higher **energy density** (ED) than other technologies offering currently. Sony partnered with several other companies first introduced lithium battery technology in 1990. Lithium ion batteries are made up of compartments called cells. Each cell is composed of three core parts: an **electrolyte** (electron provider), **anode** (electron discharger), and **cathode** (electron receiver). Cathode, the positive electrode is made from chemical compound such as lithium Cobalt oxide (LiCoO2) or lithium iron phosphate (LiFePO4). The negative electrode or, anode is made up of carbon, graphite or silicon and the electrolyte varies from one type to another.

Not Lithium Battery, but a **Lithium-ion battery** (LIB)

As it goes, lithium batteries are actually known as lithium ion batteries (LIBs). Lithium as Lithium ions (Li^+) houses in the electrolyte and shuttles toward the electrodes during charge/discharge cycles. Electrolyte is a solid or liquid material loaded with positively charged lithium ions (Li^+) that allows only ions to travel through it reaching to anode or cathode. Electrolyte separates cathode from anode and acts as an effective insulating barrier. Lithium battery anode is not made of lithium, rather Li^+ charge accumulates on anode. We will use the term LIBs throughout this book to represent this energy cell's properties. All LIB batteries work in identical manner.

During charging, positively charged lithium ions in the electrolyte are attracted to the negatively charged anode and the lithium accumulates on the anode.

CHARGE-DISCHARGE STATE of LIB

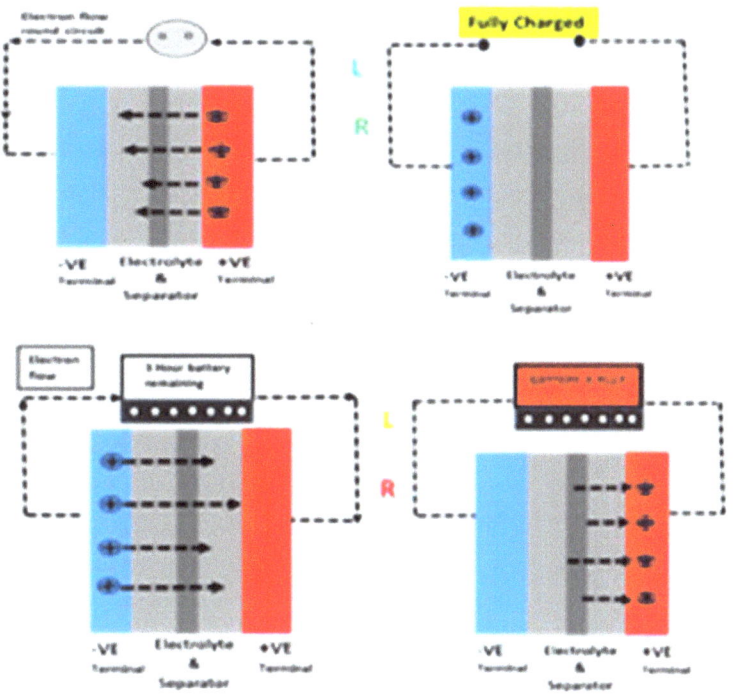

Top L) During charging time, lithium ions flow from the positive electrode to the negative electrode through the electrolyte. Electrons tend to flow in the opposite direction around the outer circuit.

Top R) When all the ions stop flowing, the battery is supposed to be fully charged and ready to work

Bottom L – Bottom R) During discharge time of the battery, the ions flow back from the negative electrode to the positive electrode. Electrons tend to flow the opposite way through the outer circuit, powering electrical devices.

2.2 LITHIUM-ION BATTERY (LIB-ALL STAGES)

Battery stores energy during the process. During discharge, lithium ions move back across the electrolyte to the positive electrode, producing the energy that powers the battery. In both cases electrons flow in the opposite direction to the ions through interconnected external circuits. Electrons do not flow through the electrolyte. If any (charged species) in the line stops moving, the reverse flow of all charge or electron circulation stops simultaneously.

10.1 Magnesium to replace Lithium in Lithium-ion battery: Professor Liwen Wan and David Prendergast of the Lawrence Berkeley National Laboratory in California conducted computer simulations experiment that demonstrated that Magnesium (Mg) can replace Lithium with little to no change in reactivity. Magnesium ions in the batteries' electrolytes can transmit electricity while **carrying double positive charge** (Mg^{+2}), this increases the device's energy density. A research of the <u>online publication Energy Trends</u>, a new technology uses electrodes made of magnesium membranes and magnesium powder and the research group reported about scope of Mg-Li swap.

10.2 A group of five research organizations, such as Pacific Northwest National Laboratory (PNNL), Harbin Institute of Technology (HIT), Wuhan University (WH), Tianjin Institute of Power Sources (TIPS), and the US Army Research Laboratory jointly conducted effectiveness of electrolyte additive called Cesium hexafluorophosphate (**$CsPF_6$**) on Lithium. $CsPF_6$ is an additive able to promote the dendrite-free growth of films made of tightly packed lithium nanorods to improve the performance of the battery and protect the negative electrode. The combined effect of Cs^+ additive and the SEI (Solid Electrolyte Interphase) contributed to lithium ions to be smoothly deposited on the surface free of dendrites, during charging. Overall, this research findings show $CsPF_6$ promotes ordered and smooth growth of lithium metal films, protects the anode, and improves battery performance.

10.3 Li-Na swap: To replace Lithium (Li) metal with Natrium (Na), researchers considered some fundamental properties of sodium. Sodium and lithium have identical tendencies to lose one electron—as measured by their electrochemical potential—which makes those good anode materials. One disadvantage of <u>sodium ions</u> (Na^+) over lithium ions is that Na_+ is nearly 25% larger than lithium ions (Li^+). The larger size makes it more difficult for sodium ions to be inserted into the crystal structure of the electrodes, where the chemical reactions take place. As a result, the ions can't move as fast, which is the root of the slow charge/discharge problem. To keep sodium ions size intact, researchers improved the efficiency of inserting Na^+ ions into the electrodes using a realistic molecular design strategy. With the flow of current density (1 A/g), the new battery delivers a reversible capacity of 160 mAh/g, which is one of the highest values reported for both inorganic Na-ion and Li-ion batteries to date. The battery also exhibits good retention capacity (70% after 400 cycles).

10.4 Lithium-ion-sulfur (Li-S) battery was synthesized by adding polysulfide to the polymer membranes of cathode. This synthesis lessened the cathode dissolution during cycling. This improves the room-temperature ionic conductivity of the polymer electrolyte. Using Li_2S_8 (EC-CDC) as the plasticizer substance to the carbonate membrane, the cell consists of i) sulfur-carbon cathode, ii) a Li-Sn-C nanostructured anode and iii) a PEO-based polysulfide-as electrolyte demonstrated good electrochemical performances in terms of stability and capacity.

Polymer electrolyte and a Li-Sn-C nanostructured alloy to replace lithium metal is key part in this new storage design. The cell is expected to guarantee high safety with advanced energy storage system.

10.5 A new study has been conducted to upgrade long-lasting, higher-capacity lithium rechargeable battery properties to satisfy common consumers' desire. The study reported that using a coating (**allucone**) on more durable silicon electrode makes it ready for a replacement for lower-capacity graphite electrodes. The US Department of Energy's Pacific Northwest National Laboratory (PNNL) research group has accomplished the study.

The coating is a naturally grown rubbery coating on nanoparticles, first thought was from silicon oxide coating. Ban's research group—developed this new coating material for <u>silicon electrodes</u>, they named it as **alucone**, Alucone coating softens the silicon particles, making it easier to expand and shrink with lithium.

X-X

2.3 CAPACITOR AS ENERGY STORAGE

CAPACITORS:

The fundamental difference between battery and capacitor is: <u>Batteries store energy in a chemical reaction, whereas capacitors store energy in an electric field.</u> Both batteries and capacitors can store strengths of different power of energy.

Capacitors can be used as like batteries, but capacitors are not battery and created to follow different mechanism. Capacitors store energy is in an electric field, whereas battery uses chemical reactions played by its constituent's electrodes and electrolytes. A basic capacitor consists of two metal plates, or conductors, separated by an insulator, such as air or a film made of plastic or ceramic. When electrons are allowed to flow from one side to another, we get electricity from the capacitor. Capacitors can discharge and charge much more quickly. Also, capacitors don't degrade over a few hundred charging cycles.

Super capacitors: Supercapacitors have two metal plates, like capacitors, but they're usually coated with a sponge like, porous material known as activated carbon (—carbon nanotubes or graphene are ideal because they are so thin). The porous material makes the surface area a lot larger and boosts the energy density. This porous films serve as supercapacitors, which store energy quickly as static charge and release it in a burst. Though they don't store as much energy as an electrochemical battery, they have long lifespans and are in wide use because they can deliver far more power than a battery. Sometimes, the capacitors also immersed in an electrolyte made of positive and negative ions dissolved in a solvent. Toyota's **hybrid racecar (TS040)** actually employs both battery, and a supercapacitor to add 480 horsepower. Therefore, Batteries and capacitors both serves different role and complements to each other.

Ultracapacitors: Ultracapacitors could play an interesting role in ultra-thin battery-powered devices such as gadgets. One Singapore research team reported of developing an ultracapacitor based on graphene and carbon nanotubes that woven into wearable devices.

Public transportation such as hybrid buses (especially in China) uses ultracapacitor technology equipped with stop-start engines. Engines turned off when the vehicle stops. It reduces the load on the battery itself. Maxwell Technologies is a supplier of start-stop engine having ultracapacitors. Each ultracapacitor's carbon electrodes end up having two layers of charge coating on its surface. However, ultracapacitors and super capacitors are often used interchangeably to mean the same device — (these are double-layered capacitor).

Capacitor Research

1. A new two-dimensional nanoporous thin film of <u>molybdenum disulfide</u> was synthesized by Rice Lab Chemist James Tour with Yang Yang, Huilong Fei and their colleagues.

Molybdenum disulfide (S-Mo-S) with three distinct layers with sulfur atoms in their own planes above and below molybdenum. The material is highly porous, short, 5- to 6-nanometer planes with a lot of edge better for catalyzing hydrogen or storage energy according to Professor Tour. The films can also serve as super capacitors to store energy quickly as static charge and release it in a burst. They have long lifespans and are in wide use for to deliver far more power than a battery.

2. **Graphene** 3D Lab Inc. Toronto, Canada (**TSX-V: GGG, OTCQB: GPHBF**) obtained a patent from U.S. Patent and Trademark Office for recent 3D printable batteries. The graphene-studded materials that have been developed by the company allow the 3D printing process to be used to fabricate a functioning battery for 3D printed objects. The company's 3D printed graphene battery is more efficient than a conventional battery because of its shape, size and specifications that it can be freely adjusted to fit a particular design of the device. A link outlining details of the 3D printed battery development is currently available on the company's website.

2.1) **i)** By exploiting Graphene's chemical property (one atom-thick form of pure carbon layer) such as: graphene's hyper-conductivity, fast charging and sensitivity to light properties UCLA researchers demonstrated that graphene can be use to store electric energy. **ii)**Graphene Quantum Dots (**GQDs**) permits the chemistry of oxygen reduction, one of the two required reactions for operation of a fuel cell according to Professor Tour of MIT. **iii)** Graphene based super capacitors developed an energy density that almost reached to that of lithium-ion batteries. Scientists from Manchester University in the U.K. predicted that a thin membrane of graphene could be use to separate hydrogen gas from air using electricity providing clean source of hydrogen to power fuel-cell cars. **iv)** At the Institute of Photonic Sciences in Spain, researchers claim graphene can produce multiple electrons for each photon of sunlight that it absorbs – generating proportionally more electricity than the 1:1 ratio of photons to electrons limited to silicon cells.

X-X

CHAPTER-3

UNITS AND ESSENTIALCALCULATIONS

LIB - OPERATION DETAILS:

LIB is a rechargeable battery made up of compartments called cells. Each cell is composed of three core parts within which: lithium ions (Li^+) move from the negative electrode to the positive electrode during discharge and reverse the course (from positive electrode to negative end) upon charging. Negative pole/electrode is the anode from which electrons flow out – enter into a power-hungry device, positive pole/electrode is the cathode where electrons re-enter to the battery once they have traveled through the circuit. Separating them is an electrolyte, a solid or liquid loaded with positively charged lithium ions that travel between the anode and cathode. Conventionally produced lithium-ion battery anodes are made of using copper foil coated with a mixture of graphite, a conductive additive, and a polymer binder. Anode of a lithium ion battery is made of graphite or silicon though pure Li as anode has been predicted to be the best.

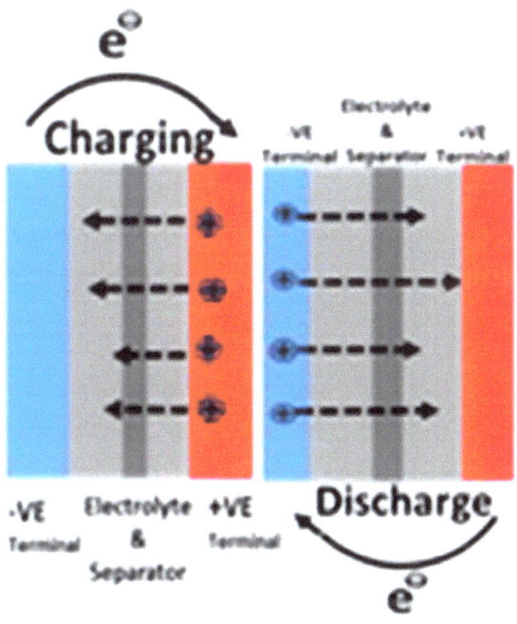

3.1: LIB in Action

LIB batteries use an intercalated lithium compound as electrode material known as the electrolyte (serves two tasks; allows ionic movement through the compartment but shields electrons passing through it.) Lithium ions (Li^+) generates, as cathodes dissociated from a variety of lithium molecules such as lithium cobalt oxide **($LiCoO_2$),** lithium manganese oxide **($LiMn_2O_4$),** or, lithium nickel oxide **($LiNiO_2$)** with the usual anode. Chemistry of LIB varies with the make-up compositions of the cells. The table below highlighted name with chemical composition of several make-up materials and their role:

TABLE-1:

Properties of Lithium Compounds in Cells

Chemical Composition	FORMULAS	Characteristic Properties
Lithium cobalt oxide	LiCoO2	High energy density with safety risks
Lithium iron-phosphate	LiFePO4	Offer lower energy density Longer lives with safety.
Lithium manganese oxide	LMO	
Lithium nickel manganese cobalt	NMC	
Lithium nickel cobalt aluminum oxide	NCA	Specialty designs aimed at particular roles
Lithium Titanate	LTO	
Lithium Sulphur	LiS	Highest performance to weight ratio

LIB Efficiency —

Pure Lithium metal as Anode:

To enhance LIB efficiency, battery manufacturers' worked to switch to anode to be made from pure Li metal, but all effort seems unsuccessful till now because manufacturers' stumbles on building such type of anode pole due to following 3 reasons:

a) Lithium's expansion during charging is "virtually infinite," and uneven relative to other materials causing cracks to form in the outer surface of the anode. Precious lithium ions to escape forming hair like dendrites. This, in turn, causes short circuit in the battery and shorten its life.

b) 2nd: Lithium anode is highly chemically reactive with the electrolyte. It uses up the electrolyte and reduces battery life.

c) 3rd: Anode and electrolyte produce heat when they come into contact. As such safety become a serious concern for Lithium only batteries.

Pure lithium anode for better efficiency is an expensive endeavor. A quest to do so has been carrying on for decades with no sign of success.

LIB- General Characteristics

LIB battery characteristics are tabulated below Property such as energy density (ED), Life-cycle, Specific capacity (Cp), charging time, frequency of charging, cost etc.

TABLE 2: LIB Properties

#	Battery properties	Advantages	Drawbacks
1	Cost	Lithium is pretty costly for large scale applications on the utility grid	X
2	Mass	Lighter than other rechargeable batteries	X
3	Power	Li-ion chemistry delivers a high open circuit voltage	X
4	Discharge rate	Low self-discharge rate (~1.5% per month)	X
5	Memory effect	Do not suffer any memory effect	X
6	Environmental effect	Rechargeable, reduced toxic landfill	X
7	Cycle life		Poor, especially for high current applications
8	Aging effect		Internal resistance rises with cycling and age
9	Safety concern		Unsafe, causes explosion, pure lithium is chemically reactive
10	Energy density (ED)	Greater, 233kwh/g (for LiFePO4)	X
11	Power Density (PD)	High (due to lightweight)	X
12	(mAh) Capacity	High, 170mAh/g for LiFePO4), depends on the electrolyte composition	X
13	Operating T	Batteries ordinarily can't operate properly at temperatures below 5 degrees Fahrenheit.	X
14	Design flexibility	Possible, helps to attain long-life span for all hand held electronics and Electric Vehicles	X
15	Expansion		Lithium expands, causing damage to anode
16	Heat generation		Anode & electrolyte produces heat, damage cell

ANODIZATION:

Anodization is an electrochemical process done to serve many purposes but traditionally employed to thicken natural oxide layers on metals. There are different approach of anodization process, for example, taking arrays of substrate sheets standing on edges and to grow the oxide film. A new approach is to grow a porous oxide film on top of the same substrate through room-temperature anodization process. (Growing a thin film of molybdenum oxide on molybdenum substrate at room temperature. This film was then exposed to sulfur vapor at 300^0C (572^0F) for one hour.

The converted material is molybdenum disulfide, no damage to its nanoporous sponge like structure (MoS_2). The Rice University research professor, James Tour said recent product, "We see anodization as a route to materials for multiple platforms in the next generation of alternative energy devices."

Battery Performances

POWER (W) = Voltage (V) x Ampere (A)

ENERGY (W-h): Amount of power a device consume or, supply in one hour.

Any battery's performance is the tradeoff between charge/discharge rates and its capacity. Electrical charge per unit weight of a battery is specific capacity.

Current density is also a measure of charge/discharge rate (C/D). For example, for a cell with charge over discharge rate is; 10A/g /10mA/g =1000 times difference.

In addition, if the cell's capacity is 72 mAh/g, then the battery's performance is poor.

Any battery's runtime depends on the efficient use of battery power that it possesses. Battery is always neutral. So, batteries need to be selected according to the application. A 10-hour application will require a different battery than does a 10-seconds frequency regulation application.

Ampere-hours (Ah) and Milliamp-hrs (mAh) units: Specific Energy is Capacity. Capacity represents specific energy in ampere-hours (Ah). Ah is the discharge current a battery can deliver over time. The unit for measuring electric charge (power) in a battery over time is milli-amp hour (shortened to: mAh). The unit is also used for to describe total amount of energy a battery can store. mA (milliampere) is 1000th of an ampere. By installing a higher Ah battery to get a longer run time; or by installing a downsize pack to get shorter run time is a design friendly option in some circuits. If a battery rated as 1500 mAh; then, it would be able to power a device drawing 100 milliamps current for 15 hours (100mA*15h=1500 mAh) and so on. A powerful battery with 3000 mAh battery be able to power the same device drawing 100 milliamps current for 30 hours (100mA*30h=3000 mAh).

Relationship-1:

To deliver 1mA current for 1hour from a battery yields, ~ 1mA*1h =1 mAh (power)

We estimated a battery life, by dividing its power (mAh) by nominal load current (mA) that a device can draw over unit time. For example, an 1800 mAh battery when connect to a 20 mA load, the approximate run time for this battery is: 1800 mA.h/20 mA = 90 hours. Likewise, the difference between a 3000 mAh and a 1000 mAh battery is, if the batteries placed in the same circuit/application a 3times run time variation is the general expectation here. In general, higher current needing device drain a battery faster.

Current measured in Amperes, is already a rate (1C/s). **By definition, Ampere- is the rate of electric charge transferred by a steady flow of current for unit time (s). We estimate the flow for one hour (3600s), by multiplying current (A) by time (s) (knocking out time units) yields charge unit only.** Here is the relationship:

Relationship-2:

1 ampere = 1C/s; 1mA = 1.10^{-3}C/s

1 mAh = $(1-10^{-3}$ C/s) *h. =

$(1-10^{-3}$ C/s)*(3600s) = 3.6 C

So, 1mAh= 3.6Coulomb, or,

3.6C = 1mAh

1C = 1mAh/3.6

Molar charge Capacity of LiFePO$_4$ (LFP):

$$LiFePO_4 + yC \xrightarrow{CHARGE} Li_{(1-x)}FePO_4 + Li_xC_y \quad \text{Equation-1}$$

$$YC + xLi + xe^- \xrightarrow{DISCHARGE} Li_xC_y \quad \text{Equation-2}$$

Rxn-2: Li-ironphosphate Molar Capc. Calc

Lithium iron phosphate (LiFePO$_4$), molar mass = 158gm. 1mole of LiFePO$_4$ =158gm

Based on the reaction set up, each molecule of lithium iron phosphate molecule yields one Li^+ ion. Therefore, 1mole of $LiFePO_4$ molecules will deposit →1 mole of Li^+ ions. Using Faraday's law, we get, 1 mole charge = 96,500Coulombs;

Molar charge Capacity of $LiFePO_4$ has been determined as follows,

1 mole of Li^+ charge = 96500Coulombs,

1 coulomb = 1mah/3.6 (from relationship-2),

96500C = 96500C*mAh/3.6 = 26805.55mAh charge (after Division)

Molar mass of $LiFePO_4$ =158 gm; or,

1mole of Li^+ ions = 158gm= 26805.55mAh

Molar Charge capacity ($LiFePO_4$) = 26805mAh/158g= **170mAh/gm.**

LIB Cells (Lithium Content):

Lithium is popular for car battery because of its energy density relative to weight. As Tesla and other EV makers have demonstrated, one can pack a lot of KWh- into a relatively small space. That's why lithium battery is good for electric cars.

Energy density is a measure of how much energy a battery can store, in a given size or mass. So, a battery with a higher energy density can power a load for longer time than with a low energy density battery of identical physical size or mass. Its unit: Wh/Kg, or, Wh/m^3. [Energy (watt-hours)].

Specific energy density defines battery **capacity** in weight or, by volume (Wh/Kg), or (Wh/l). For **long run time** with **moderate load** uses **high specific energy** to keep the electric device in charged state for longer hours than with a low ED battery Means, lighter the weight of the container, higher the ED of a battery. ED is critical property of a battery.

TABLE 3:

ELECTRON DENSITY (ED) Values

Cell Type Name	Formula	ED Value
Zinc Bromide	ZnBr2	70Wh/l
Zinc Polyiodides	ZnI3 (PNNL Research)	167Wh/l
Lithium ion	Li	18Wh/l
Nickel Metal Hydride	NiMH	140Wh/Kg
Lithium iron Phosphate	Li(Fe(PO4)3	233Wh/l

Power Density (PD) definition: The ability of a battery to deliver power (fixed VA) per pound /(gm) of battery weight. Units are W/Kg or, W/m^3. PD depends upon the construction design of an energy cell. **A higher power density (PD) means the battery can deliver higher than normal power for its size and weight.** Batteries with high power density will charge faster than low PD battery. Consider a Hybrid Bus, it has one battery for starting the engine, and another one to run the onboard electrical lines such as interior lights etc.

• The starting battery needs to deliver a high power to crank the engine, but not for very long. Therefore it needs to be built with a high power density, but not with a (high energy density) as the starter motor is only engaged for few secs only.

• The interior battery, in contrast is expected to deliver its energy over a much longer period of time-from few hours to few days. Current draw is much less. So, energy density for this battery needs to be high, but power density to be low. By knowing the current demand of a device which is its power consumption, power density or, energy density factors can be used to determine required battery size for any application.

Each cell of LIB battery requires ~ **0.3 gms** of **lithium metal** to produce 1 ampere-hour (Ah) of power. If a battery is rated as 2500 mAh per cell and it contain 6 cells, total lithium (gm) needed for this battery to produce enough power is: 2.5Ah*0.3gm*6=4.5gms of Li.

Motor industries standard research maintains theoretical rule of: 1 kilogram of lithium to enable a 6 kWh battery, which is in line with the productions. The table below highlighted amount of lithium required per battery by the sales figures for each specific vehicle to get an estimate for the total amount of lithium consumed by specific automakers.

TABLE-4: Lithium Estimation in Kg unit

Amount of Lithium Present in Electric Vehicles

Electric Vehicle Models	Battery (kWh)	Lithium Required (kg)
Standard Value	6.0	1.0
Chevy Volt (plug-in-hybrid)	16.5	2.8
Nissan Leaf	24.0	4.0
Chevy Bolt (GM)	60.0	10.0
Tesla Model S	70.0	11.66

This data justifies that more **kWh** a battery claims, further the driving range on electric power – and more lithium contained within the battery. Source:-Fox Davies Research SignumBox.

Lithium Battery for Electric Vehicles (EVs):

Each company entering EV trade must define its own role and business model, based on the chemistry and capabilities of technology it will play the game. General rule for electric vehicles is: more **'kWh'** a battery possesses, further the **mileage** it can cover with electric power – coincidentally, more lithium contained within that battery. Tesla offers vehicles with 70 kWh or 85 kWh batteries. Tesla cars use lithium ion batteries for both storing energy and providing power for the vehicles, The Nissan Leaf sports a 24 kWh battery, while the Chevy Bolt from General Motors use a 60 kWh battery. The Figure-set below summarized new LIB info.

PLUG-IN ELECTRIC VEHICLES:

Motor vehicles with a rechargeable battery that cannot be charged from electric grid are called **plug-in EVs**. Plug-in EVs are the beneficiaries from BASF research. Huge number of new cars are now being fitted with engine start-stop systems to reduce fuel consumption, and the most advanced of those systems require a bolder battery. Lithium battery would be used to store regenerative braking energy for restart when the start-stop function is called into use. And. electric cars use capacitors for acceleration purpose (quick charge- discharge function). It helps car battery to be **smaller** size and **lighter**, thereby **increasing mileage** cover by the car.

LIB Giga Factory: A Global Initiative

A global market of most commonly used lithium battery estimated first for $11bn, now predicted to grow to $60bn by 2020 is a realism. For mass production of lithium battery, companies' are now setting up Gigafactories, a production center whose annual production capacity is **GWh**. A world with **200 Gigafactories be to enable a full transition from gasoline powered vehicles to full blown electric vehicles-powered by lithium batteries-**this vision is from TESLA's CEO. Tesla Motors remained at the forefront of this initiative with other companies advancing towards the goal stepwise. Building factories and embracing ideas to introduce EV's technologies. Once shifted, that would create a pretty different world from today that we are now living in. Of course, that world would be better because a CO_2 emission free greener world. The world would require huge lithium supply. Fortunately, the world has **plenty** of reserve lithium. Even, **triple to its** lithium production from the current level and still have **135 years** of reserve supply according to US Geological (USGS) surveys. Lithium if fully recyclable. Although it's no economical today, recycling lithium batteries could one day provide cost recapture for manufacturers.

Tips to last long cellphone battery:

#Cell phone battery only has a limited amount of times it can charge. Keeping it plugged causes this number to go down significantly. This is one of the troublemakers as to why our cell phone charge keeps lasting less and less as time goes by.

keeping Away From Very High/Low Temperatures will help battery lasts long.

#By lowering device screen brightness will help battery lasts long.

By turning Off WiFi/GPS/Bluetooth when not in use increase battery life.

#And finally, being inside a poor signaling areas need to be avoided in order to increase cell phone's battery last longer.

CHAPTER-4

MISCELLANEOUS

DOMINANT ENERGY STORAGEA FACILITIES WORLDWIDE

- Randy Ross in **Pleasanton, Calif., USA** has an energy storage system which is part of a pilot project from SolarCity. He currently pays just $1 a year for the battery system. His lithium-ion battery storage system, contained in a 4-foot-tall metal box mounted on the garage walls.

- Tesla, in pursuing new energy storage market is repositioning itself as a full-energy service provider and plans to add battery-based storage systems to its solar projects. **Reno, Nevada, USA** is the Tesla Motor's Gigafactory.

Also in the race are:

- Primus Power, based in **Hayward, CA** has systems in the field, including at the Marine Corps Air Station in **Miramar, CA.**

- Sunnyvale-based EnerVault, recently dedicated its long-duration battery storage at an almond orchard in **Turlock city in CA**, USA. The battery can be charged from the orchard's solar system and then be used to provide power to pump ground water for irrigation.

- Energy Storage at Grid Scale: A123 Gets Li-Ion to Market, builds nanophosphate lithium-ion batteries and systems for the transportation, electric grid, and commercial markets. In 2009, A123 and AES began commercial operation of a 12-megawatt spinning reserve project at AES Gener's Los Andes substation in the **Atacama Desert in Chile.**

- Samsung's lithium battery division and Youcious, a German energy storage company, will work with Duke Energy, NC, USA to replace the Lead acid cells in the battery with lithium ion battery cells by the first quarter of 2016. During the transition, the new batteries will operate in combination with advanced lead-acid technology currently in use at the site. **Charlotte, NC-based Duke** installed the original battery at the 156-megawatt Notrees windmill farm in 2012.

- Green Charge Networks, a startup deploying energy storage equipment for commercial customers. Founded in 2009, Green Charge's GreenStation has been installed by

7-Eleven, Walgreens, Levi's Stadium, UPS, school campuses, and **cities across California**, including Redwood City and Lancaster.

- **PG&E, Southern California Edison** and San Diego Gas & Electric are now required by state regulators to collectively buy 1.3 gigawatts of energy storage capacity by 2020, enough to supply roughly 1 million homes.

- **Greensmith** writes software that integrates batteries with inverters and allows energy storage to work with the grid at scale. Greensmith works with a variety of battery chemistries from different vendors to multiple inverters and power electronics partners.

- **Apple**'s solar farms put the energy onto the grid and then the data center draws equivalent power from the grid. Apple didn't want to invest in energy storage tech for the solar sites, as that would bump up the cost of clean power considerably.

- British Government is providing $90m funding to Dyson to develop a new battery for electric vehicle. Sakti3 was founded 8 years ago by former University of Michigan engineering professor Ann Marie Sastry and developed what's called a "solid state" lithium-ion battery. The battery uses a solid material, rather than a liquid, making it potentially safer and less flammable. Sakti3's battery is also high performance. A couple of years ago, the startup said if its batteries were used in an electric car, like a Tesla Model S, it could double range. Indeed, Sakti3 battery cells created an electric car with a range of **480 miles**, according to the company. Price: $100 per kilowatt hour.

TABLE -5: DAILY POWER USES

iPhone (and other devices) Use?

Device Name	Other Description (if any)	Energy Use (W/h)	Cost, $	Source
iPad	Drain/charge everyday	12kWh/Y	1.50/Y	Electric Power Research Institute
Mobile phone	Holding charge-1440mAh	2kWh/Y	0.25c	EPRI and Apple
Laptop	---	72kWh/Y	8.0/Y	EPRI
Plasma TV	Big screen; av. 5hrs./day	360kWh/Y	45.0/Y	EPRI and CNET
LCD based TV	do	do	20.0/Y	do
X Box Play Sta	Draw an Av. 100W	---	40.0/Y	Hettinger, Mullins and Arenado
60W Light Bulb	90% as heat, 10h/day	220kWh/Y	26.0/Y	---
LED light Bulb	60W equivalent -10h/day	37kWh/Y	4.40/Y	---
Desktop Computer	(+modem, wireless and router)	(300kWh+90kWh)/Y	36/Y	NRDC
General Automobile	Av drive 15,000m/Y, 600G gas	25m/G, 3.60/G	2,200/Y	General Motors
TESLA-S	300mile range battery	85kWh/Y	450/Y	TESLA Engineers
Washer/Dryer	---	---	300/Y	Mr. Electricity/ Energystar.gov
Water Heater	18% of Av. Homes energy bill	---	600/Y	EPA, Mr. Electricity
Microwave	15min. Uses,0.36kWh	Variable	rate 4c	Mr. Electricity
Electric Oven	350degree for an hour-2kWh	Variable	rate 24c	Mr. Electricity
Refrigerator	A 2001 manufactured Fridge	350kWh/Y	45.0/Y	Mr. Electricity
Heating/Cooling	2.5ton ac for an hour	350kWh/Y	1000.0/Y	---
VCR (standby state)	5.0W/hour	43kWh/Y	5.0/Y	Lawrence Barkley Laboratory

X-X

ACKNOWLEDGEMENT

The author wishes to thank everyone, whose help made this book possible. This book would not have been possible without this help. My solar research experience and training on Solar Panel created extraordinary passion for energy production and storage facilities for me. That reflected on energy cells study based on what I produced this research file. I look forward to path for development.

REFERENCES

1. http://www.youtube.com/watch?feature=player_embedded&v=sIxt6nMf-IQ

2. http://www.ted.com/talks/
    donald_sadoway_the_missing_link_to_renewable_energy?language=en#t-782814

3. http://gcell.com/dye-sensitized-solar-cells; ii) Yu, M. et al. Integrating a redox-coupled dye-sensitized photoelectrode into a lithium–oxygen battery for photoassisted charging. Nat. Commun. 5:5111 doi: 10.1038/ncomms6111 (2014 [

4. http://www.bbc.com/news/magazine-27829874

5. http://www.technologyreview.com/news/535251/old-battery-type-gets-an-energy-boost/

6. https://youtu.be/ZKIcYk7E9lU

7. Chengliang Wang, et al. "Extended π-Conjugated System for Fast-Charge and -Discharge Sodium-Ion Batteries." Journal of the American Chemical Society. DOI: 10.1021/jacs.5b00336

8. Yang He, Daniela Molina Piper, MengGu, Jonathan J. Travis, Steven M. George, Se-Hee Lee, Arda Genc, Lee Pullan, Jun Liu, and Artificial Layers on Silicon Nanoparticles for Lithium Ion Batteries, ACS Nano, October 27, 2014, DOI: 10.1021/nn505523c

9. http://www.graphene3dlab.com

10. Oxygen Reduction." Huilong Fei, et al. ACS Nano, Just Accepted Manuscript, September 24, 2014. DOI: 10.1021/nn504637y

11. http://www.businessinsider.com/use-tesla-supercharger-station-darien-connecticut-2015-5

X-X

END